SpringerBriefs in Applied Sciences and Technology

SpringerBriefs in Continuum Mechanics

Series Editors

Holm Altenbach, Institut für Mechanik, Lehrstuhl für Technische Mechanik, Otto von Guericke University Magdeburg, Magdeburg, Sachsen-Anhalt, Germany

Andreas Öchsner, Faculty of Mechanical Engineering, Esslingen University of Applied Sciences, Esslingen am Neckar, Germany

These SpringerBriefs publish concise summaries of cutting-edge research and practical applications on any subject of Continuum Mechanics and Generalized Continua, including the theory of elasticity, heat conduction, thermodynamics, electromagnetic continua, as well as applied mathematics.

SpringerBriefs in Continuum Mechanics are devoted to the publication of fundamentals and applications, presenting concise summaries of cutting-edge research and practical applications across a wide spectrum of fields. Featuring compact volumes of 50 to 125 pages, the series covers a range of content from professional to academic.

More information about this subseries at https://link.springer.com/bookseries/10528

Andreas Öchsner

Micromechanics
of Fiber-Reinforced Laminae

 Springer

Andreas Öchsner
Faculty of Mechanical Engineering
Esslingen University Applied Sciences
Esslingen, Germany

ISSN 2191-530X ISSN 2191-5318 (electronic)
SpringerBriefs in Applied Sciences and Technology
ISSN 2625-1329 ISSN 2625-1337 (electronic)
SpringerBriefs in Continuum Mechanics
ISBN 978-3-030-94090-4 ISBN 978-3-030-94091-1 (eBook)
https://doi.org/10.1007/978-3-030-94091-1

This Springer imprint is published by the registered company Springer Nature Switzerland AG
The registered company address is: Gewerbestrasse 11, 6330 Cham, Switzerland

Preface

Composite materials, especially fiber-reinforced composites, are gaining increasing importance since they can overcome the limits of many structures based on classical metals. Particularly, the combination of a matrix with fibers provides far better properties than the components alone. Despite their importance, many engineering degree programs do not treat the mechanical behavior of this class of advanced structured materials in detail, at least on the Bachelor degree level. Thus, some engineers are not able to thoroughly apply and introduce these modern engineering materials in their design process.

This SpringerBriefs volume provides the first introduction to the mircomechanics of fiber-reinforced laminae, which deals with the prediction of the macroscopic mechanical lamina properties based on the mechanical properties of the constituents, i.e., fibers and matrix. The focus is on unidirectional lamina which can be described based on orthotropic constitutive equations. Three classical approaches to predict the elastic properties, i.e., the mechanics of materials approach, the elasticity solutions with contiguity after Tsai, and the Halpin-Tsai relationships, are presented. The quality of each prediction is benchmarked based on two different sets of experimental values. The volume concludes with optimized representations, which were obtained based on the least squares approach for the used experimental data sets.

Esslingen, Germany   Andreas Öchsner
November 2021

Contents

Symbols and Abbreviations

Latin Symbols (capital letters)

A area, cross-sectional area
C fiber contiguity
E modulus of elasticity
G shear modulus
K bulk modulus
N internal normal force
V volume

Latin Symbols (small letters)

a geometric dimension
     numerical coefficient
b geometric dimension
k fiber misalignment factor
m mass
t thickness

Greek Symbols (small letters)

α rotation angle
γ shear strain
ε normal strain
η coefficient (Halpin-Tsai approximations)
ν Poisson's ratio

ξ coefficient (Halpin-Tsai approximations)
ρ relative density, specific gravity
ϱ volumetric mass density
σ normal stress
τ shear stress
ϕ volume fraction
ψ mass fraction

Mathematical Symbols

Indices, Superscripted

$\ldots^{\text{upper}}$ upper bound
$\ldots^{\text{lower}}$ lower bound

Indices, Subscripted

$\ldots_1$ direction *1*
$\ldots_2$ direction *2*
$\ldots_{12}$ plane *12*
$\ldots_f$ fiber
$\ldots_m$ matrix
$\ldots_{\text{ref}}$ reference substance

Abbreviations

CLT   classical laminate theory
MMA   mechanics of materials approach
SI   International System of Units

Chapter 1
Introduction

Abstract The first chapter introduces the major concept of composite material, i.e., the combination of different components, to obtain in total much better properties than a single component for itself. The focus is on fiber-reinforced composites where a single layer has unidirectionally aligned reinforcing fibers. The second part of this chapter introduces the concept of reference numbers, which are commonly used to predict and characterize the properties of composites depending on their composition. The common approaches are related to the volume, mass, or relative density.

1.1 Preliminary Comments

Many classical engineering materials (metals) reach in advanced technical applications their limits and scientists permanently seek to improve the properties or to propose new materials. To judge the lightweight potential of a material in the elastic range, one may use the so-called specific modulus, i.e., the ratio between the Young's modulus and the mass density, see Table 1.1. One can see that a typical value for metals is around 20.

A typical concept to design materials with better performance is to compose different constituents to a so-called composite material, where the entire composite reveals better properties (sometimes evaluated as a spectrum of different properties) than any of the single constituents. Out of the various types of composite materials, fiber-reinforced plastics are a common choice for high-performance requirements in lightweight applications. A typical basic layer, the so-called lamina, can be composed of unidirectional fibers which are embedded in a matrix. In a second step, layers of laminae may be stacked under different fiber angles to a so-called laminate, which reveals—depending on the stacking sequence—different types of anisotropy/isotropy. Typical fiber and matrix properties are summarized in Tables 1.2 and 1.3. It can be seen that most of the fibers clearly outperform classical engineering metals. For example, carbon and boron fibers reveal a specific modulus above 100. Obviously, the mixing of fibers and matrix will decrease again these values to a certain extend but the superior properties of fiber-reinforced plastics compared to classical metals will remain.

A. Öchsner, *Micromechanics of Fiber-Reinforced Laminae*,
SpringerBriefs in Continuum Mechanics,
https://doi.org/10.1007/978-3-030-94091-1_1

Table 1.1 Properties of different classical engineering materials: E: Young's modulus; ν: Poisson's ratio; ϱ mass density, E/ϱ specific modulus (here calculated based on averaged density in case of density intervals). Adapted from [2, 5]

	E, $\frac{\text{N}}{\text{mm}^2}$	ν, $-$	ϱ, $10^{-6}\ \frac{\text{kg}}{\text{mm}^3}$	$\frac{E}{\varrho}$, $\frac{10^9\,\text{N}\,\text{mm}}{\text{kg}}$
Iron	196000	0.293	7.9	24.81
Iron-based super-alloys	193000–214000		7.9–8.3	23.83-26.42
Ferritic steels, low-alloy	196000–207000	0.27–0.3	7.8-7.85	25.05–26.45
Stainless austenitic steels	190000–200000	0.25-0.3	7.5–8.1	24.36-25.64
Mild steel	200000	0.27–0.3	7.8-7.85	25.56
Cast irons	170000–190000	0.26-0.275	6.9–7.8	23.13–25.85
Aluminum	69000	0.345	2.7	25.56
Aluminum alloys	69000–79000	0.32-0.40	2.6-2.9	25.09–28.73
Titanium	116000	0.361	4.5	25.78
Titanium alloys	80000–130000		4.3–5.1	17.02–27.66
Magnesium	44700	0.291	1.74	25.69
Magnesium alloys	41000–45000		1.74-1.88	22.65-24.86

Table 1.2 Properties of typical fibers: E: Young's modulus; ν: Poisson's ratio; ϱ mass density, E/ϱ specific modulus. Adapted from [3, 4, 6]

	E, $\frac{\text{N}}{\text{mm}^2}$	ν, $-$	ϱ, $10^{-6}\ \frac{\text{kg}}{\text{mm}^3}$	$\frac{E}{\varrho}$, $\frac{10^9\,\text{N}\,\text{mm}}{\text{kg}}$
E-glass	74000	0.25	2.6	28.46
R-glass	86000	0.2	2.5	34.4
Carbon (high modulus)	390000	0.35	1.8	216.67
Carbon (high strength)	230000	0.3	1.75	131.43
Aramid (Kevlar®)	130000	0.4	1.45	89.66
Boron	400000	0.2	2.6	153.85

Table 1.3 Properties of typical matrices: E: Young's modulus; ν: Poisson's ratio; ϱ mass density, E/ϱ specific modulus. Adapted from [3, 6, 9]

	$E, \frac{\text{N}}{\text{mm}^2}$	$\nu, -$	$\varrho, 10^{-6} \frac{\text{kg}}{\text{mm}^3}$	$\frac{E}{\varrho}, \frac{10^9 \,\text{N mm}}{\text{kg}}$
Thermosets				
Epoxy	4500	0.4	1.2	3.75
Polyester	4000	0.4	1.2	3.33
Polycarbonate	2400	0.35	1.2	2.00
Thermoplastics				
Polypropylene (PP)	1200	0.4	0.9	1.33
Polyamide (PA)	2000	0.35	1.1	1.82
Polyetheretherketone (PEEK)	4000	0.38	1.3	3.08

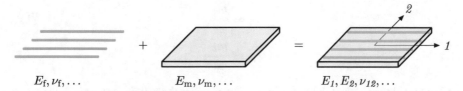

$$E_f, \nu_f, \ldots \qquad E_m, \nu_m, \ldots \qquad E_1, E_2, \nu_{12}, \ldots$$

Fig. 1.1 Schematic illustration of the micromechanics concept (index 'f': fiber; index 'm': matrix; index '1,2': lamina)

It should be noted here in regards to Table 1.2 that some fibers do not show isotropic properties, i.e. the properties in axial and transverse direction differ significantly, for example, in the case of carbon and aramid, see [7].

The mechanical description of laminae can be divided in the so-called maromechanics and the micromechanics of laminae. The macromechanics of fiber-reinforced laminae covers the mechanical response under a macroscopic external load. In this context, a stress and/or strain analysis is done and the failure may be predicted, see [1, 8]. The micromechanics of laminae predicts the macroscopic mechanical lamina properties based on the mechanical properties of the constituents, i.e., fibers and matrix, see Fig. 1.1. These macroscopic properties (such as E_1, E_2, ...) are commonly provided as a function of some fractions, for example the fiber volume fraction, see Sect. 1.2.

1.2 Reference Numbers to Characterize Composition

As mentioned in the previous section, macroscopic lamina properties are commonly provided as functions of some fractions. This allows to estimate the property based on given volume or weight/mass fractions. The fiber and matrix *volume* fractions are defined as

$$\phi_f = \frac{V_f}{V},$$ (1.1)

$$\phi_m = \frac{V_m}{V},$$ (1.2)

where V_f is the volume of the fibers, V_m is the volume of the matrix, and V is the volume of the entire composite. The following relationships hold between the volume fractions and the volumes themselves:

$$\phi_f + \phi_m = 1 \quad \text{and} \quad V_f + V_f = V.$$ (1.3)

Fiber and matrix *mass* fractions are defined as

$$\psi_f = \frac{m_f}{m},$$ (1.4)

$$\psi_m = \frac{m_m}{m},$$ (1.5)

where m_f is the mass of the fibers, m_m is the mass of the matrix, and m is the mass of the entire composite. The following relationships hold between the mass fractions and the masses themselves:

$$\psi_f + \psi_m = 1 \quad \text{and} \quad m_f + m_f = m.$$ (1.6)

One may find also in scientific literature the so-called relative density (or sometimes called the specific gravity), which is defined as

$$\rho_f = \frac{\varrho_f}{\varrho_{ref}},$$ (1.7)

$$\rho_m = \frac{\varrho_m}{\varrho_{ref}},$$ (1.8)

where $\varrho_f = m_f / V_f$ is the mass density of the fibers, $\varrho_m = m_m / V_m$ is the mass density of the matrix, and ϱ_{ref} is the mass density of a reference substance.[1]
The relationship between the volume and mass fractions can be expressed as follows [1, 10]:

---

[1] If the reference is not explicitly stated then it is normally assumed to be water (H_2O) at 3.98°C, which is the temperature at which water reaches its maximum density: $\varrho_{H_2O}(3.98°C) = 999.95\,kg/m^3 \approx 1000\,kg/m^3$ or $\varrho_{H_2O}(3.98°C) = 0.99995\,g/cm^3 \approx 1.0\,g/cm^3$.

$$\phi_m = \frac{\dfrac{\varrho_f}{\varrho_m}}{\dfrac{1}{\psi_m} - 1 + \dfrac{\varrho_f}{\varrho_f}} = \frac{\dfrac{\rho_f}{\rho_m}}{\dfrac{1}{\psi_m} - 1 + \dfrac{\rho_f}{\rho_f}}, \tag{1.9}$$

or

$$\phi_f = 1 - \frac{\dfrac{\varrho_f}{\varrho_m}}{\dfrac{1}{\psi_m} - 1 + \dfrac{\varrho_f}{\varrho_f}} = 1 - \frac{\dfrac{\rho_f}{\rho_m}}{\dfrac{1}{\psi_m} - 1 + \dfrac{\rho_f}{\rho_f}}. \tag{1.10}$$

The following sections contain many graphical representaions of mechanical properties and the common approach, i.e., the properties are plotted over the fiber volume fraction (ϕ_f), is followed.

References

1. Altenbach H, Altenbach J, Kissing W (2018) Mechanics of composite structural elements. Springer, Singapore
2. Ashby MF, Jones DRH (2005) Engineering materials 1: an introduction to properties, applications and design. Elsevier, Amsterdam
3. Berthelot J-M (1999) Composite materials: mechanical behavior and structural analysis. Springer, Berlin
4. Chamis CC (1970) Characterization and design mechanics for fiber-reinforced metals. NASA Technical Note NASA TN D-5784. National Aeronautics and Space Administration, Washington
5. Gale WF, Totemeir TC (eds) (2004) Smithells metals reference book. Elsevier Butterworth-Heinemann, Burlington
6. Gay D, Hoa SV, Tsai SW (2003) Composite materials: design and application. CRC Press, Boca Raton
7. Kaw AK (1997) Mechanics of composite materials. CRC Press, Boca Raton
8. Öchsner A (2021) Foundations of classical laminate theory. Springer, Cham
9. Rae PJ, Brown EN, Orler EB (2007) The mechanical properties of poly(ether-ether-ketone) (PEEK) with emphasis on the large compressive strain response. Polymer 48:598–615
10. Tsai SW, Springer GS, Schultz AB (1963) The composite behavior of filament-wound materials. In: Fourteenth International Astronautical Congress, Paris, France, September 1963, Paper No. 139

Chapter 2
Prediction of Elastic Properties of Laminae

Abstract This chapter introduces three classical theories to predict the macroscopic mechanical lamina properties based on the mechanical properties of the constituents, i.e., fibers and matrix. The focus is on unidirectional lamina which can be described based on orthotropic constitutive equations. In detail, predictions for the modulus of elasticity in and transverse to the fiber direction, the major Poisson's ratio, as wells as the in-plane shear modulus are provided. The mechanics of materials approach, the elasticity solutions with contiguity after Tsai, and the Halpin-Tsai relationships, are presented.

2.1 Mechanics of Materials Approach

The prediction of elastic constants for two-phase materials was already a long time before fiber-reinforced laminae discussed in scientific literature. These early predictions were related, for example, to a metal with finely dispersed particles of an alloying material which has higher mechanical properties than the base material [6].

The following sections will repeat the derivations of these classical approximations, which are known as the mechanics of materials approach. All these derivations are made under the assumption that each constituent, i.e., the fiber and the matrix, behaves itself as an isotropic material.

2.1.1 Modulus of Elasticity in Fiber Direction

To estimate the modulus of elasticity in fiber direction (E_1), it is assumed that the strains of the fiber ($\varepsilon_{f,1}$), matrix ($\varepsilon_{m,1}$) as well as the entire composite (ε_1) are the same in the loading direction, see Fig. 2.1 for schematic representations of the constituents in parallel arrangement.

© The Author(s), under exclusive license to Springer Nature Switzerland AG 2022   7
A. Öchsner, *Micromechanics of Fiber-Reinforced Laminae*,
SpringerBriefs in Continuum Mechanics,
https://doi.org/10.1007/978-3-030-94091-1_2

(a)   **(b)**

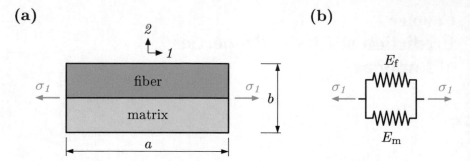

Fig. 2.1 Mechanical models to calculate the elastic modulus in fiber direction: **a** two-phase representation, **b** springs-in-parallel representation

This equal strain condition is expressed as:

$$\varepsilon_1 = \varepsilon_{f,1} = \varepsilon_{m,1} = \frac{\Delta a}{a}. \tag{2.1}$$

The stress in each constituent, as well as in the entire composite, can be expressed as:

$$\sigma_{f,1} = E_f \varepsilon_{f,1} \overset{(2.1)}{=} E_f \varepsilon_1, \tag{2.2}$$

$$\sigma_{m,1} = E_m \varepsilon_{m,1} \overset{(2.1)}{=} E_m \varepsilon_1, \tag{2.3}$$

$$\sigma_1 = E_1 \varepsilon_1. \tag{2.4}$$

Considering the internal normal force (N), the force equilibrium in fiber (or loading) direction yields:

$$N_1 = \sigma_1 A = N_{f,1} + N_{m,1} = \sigma_{f,1} A_f + \sigma_{m,1} A_m, \tag{2.5}$$

which yields under the assumption of a constant thickness in Fig. 2.1a and under consideration of Eqs. (2.2)–(2.3) the following expression:

$$E_1 \varepsilon_1 = E_f \varepsilon_1 \frac{A_f}{A} + E_m \varepsilon_1 \frac{A_m}{A} \tag{2.6}$$

$$= E_f \varepsilon_1 \phi_f + E_m \phi_1 \phi_m, \tag{2.7}$$

or finally under consideration of Eq. (1.3):

$$E_1 = E_f \phi_f + E_m (1 - \phi_f). \tag{2.8}$$

This estimation is also known as the rule of mixtures or Voigt rule.

Fig. 2.2 Modulus of elasticity in fiber direction as a function of the fiber volume fraction (MMA)

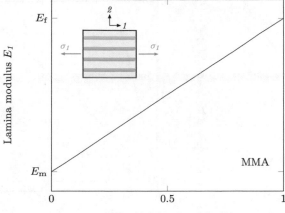

A graphical representation of the estimation for the modulus of elasticity in fiber direction is given in Fig. 2.2. A linear transition between the modulus of the matrix and the modulus of the fibers is obtained.

Equation (2.8) can be normalized either by the modulus of the matrix or the modulus of the fibers to obtain the following normalized representations of the modulus in fiber direction:

$$\frac{E_1}{E_m} = \frac{E_f}{E_m}\phi_f + (1 - \phi_f), \tag{2.9}$$

$$\frac{E_1}{E_f} = \phi_f + \frac{E_m}{E_f}(1 - \phi_f), \tag{2.10}$$

which allows a graphical representation of the modulus of elasticity in fiber direction for different ratios of the moduli, see Fig. 2.3.

2.1.2 Modulus of Elasticity Transverse to Fiber Direction

To estimate the modulus of elasticity transverse to the fiber direction (E_2), it is assumed that the stresses of the fiber $(\sigma_{f,2})$, matrix $(\sigma_{m,2})$ as well as the entire composite (σ_2) are the same in the loading direction, see Fig. 2.4 for schematic representations of the constituents in series arrangement.

This equal stress condition is expressed as:

$$\sigma_2 = \sigma_{f,2} = \sigma_{m,2} = \varepsilon_2 E_2 = \varepsilon_{f,2} E_f = \varepsilon_{m,2} E_m. \tag{2.11}$$

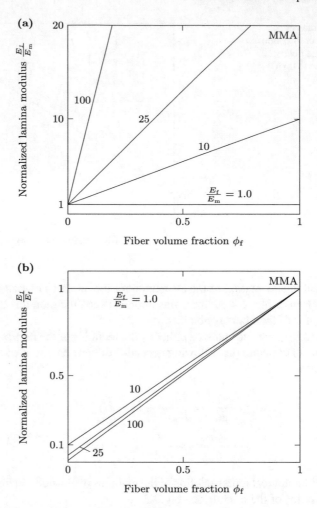

Fig. 2.3 Normalized modulus of elasticity in fiber direction as a function of the fiber volume fraction (MMA): **a** normalized by matrix modulus, **b** normalized by fiber modulus

The total deformation in loading direction is the sum of the deformations of both constituents:

$$\Delta b = \Delta b_\mathrm{f} + \Delta b_\mathrm{m}\,, \tag{2.12}$$

or expressed in terms of strains:

$$\frac{\Delta b}{b} = \varepsilon_2 = \frac{\Delta b_\mathrm{f}}{b} + \frac{\Delta b_\mathrm{m}}{b}\,. \tag{2.13}$$

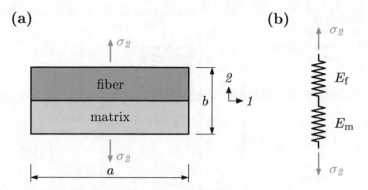

Fig. 2.4 Mechanical models to calculate the elastic modulus transverse to the fiber direction: **a** two-phase representation, **b** springs-in-series representation

Considering the volume fractions, Eq. (2.13) can be written as:

$$\varepsilon_2 = \frac{\Delta b_f}{\phi_f b} \times \frac{\phi_f b}{b} + \frac{\Delta b_m}{\phi_m b} \times \frac{\phi_m b}{b}, \tag{2.14}$$

which yields under consideration of[1] $\phi_f b = b_f$ and $\phi_m b = b_m$:

$$\varepsilon_2 = \frac{\Delta b_f}{b_f} \phi_f + \frac{\Delta b_m}{b_m} \phi_m = \varepsilon_{f,2} \, \phi_f + \varepsilon_{m,2} \, \phi_m. \tag{2.15}$$

Introducing the equal stress condition (2.11) in the last relationship gives:

$$\varepsilon_2 = \frac{\sigma_2}{E_2} = \frac{\sigma_2}{E_f} \phi_f + \frac{\sigma_2}{E_m} (1 - \phi_m), \tag{2.16}$$

or

$$\frac{1}{E_2} = \frac{\phi_f}{E_f} + \frac{1 - \phi_m}{E_m}, \tag{2.17}$$

or finally as:

$$E_2 = \frac{E_f E_m}{E_m \phi_f + E_f (1 - \phi_f)}. \tag{2.18}$$

This estimation is also known as the inverse rule of mixtures or Reuss rule.

A graphical representation of the estimation for the modulus of elasticity transverse to the fiber direction is given in Fig. 2.5. A nonlinear transition between the modulus of the matrix and the modulus of the fibers is obtained.

---

[1] This assumes for the width $a = a_f = a_m$ and for the thickness $t = t_f = t_m$. Strictly speaking, this would be the case if Poisson's ratios of the fiber and the matrix would be the same.

Fig. 2.5 Modulus of elasticity transverse to fiber direction as a function of the fiber volume fraction (MMA)

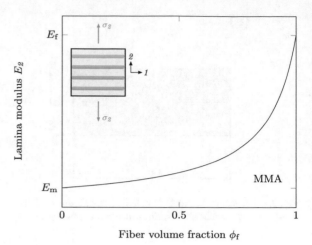

Equation (2.18) can be normalized either by the modulus of the matrix or the modulus of the fibers to obtain the following normalized representations of the modulus transverse to the fiber direction:

$$\frac{E_2}{E_m} = \frac{1}{\frac{E_m}{E_f} \phi_f + (1 - \phi_f)}, \tag{2.19}$$

$$\frac{E_2}{E_f} = \frac{1}{\phi_f + \frac{E_f}{E_m}(1 - \phi_f)}, \tag{2.20}$$

which allows a graphical representation of the modulus of elasticity transverse to the fiber direction for different ratios of the moduli, see Fig. 2.6.

2.1.3 Major Poisson's Ratio

To estimate the major Poisson's ratio (ν_{12}), it is assumed that the strains of the fiber ($\varepsilon_{f,1}$), matrix ($\varepsilon_{m,1}$) as well as the entire composite (ε_1) are the same in the loading direction, see Fig. 2.7 for schematic representations of the problem.

This equal strain condition is expressed as:

$$\varepsilon_1 = \varepsilon_{f,1} = \varepsilon_{m,1}. \tag{2.21}$$

The total deformation perpendicular to the loading direction is the sum of the deformations of both constituents:

Fig. 2.6 Normalized modulus of elasticity transverse to fiber direction as a function of the fiber volume fraction (MMA): **a** normalized by matrix modulus, **b** normalized by fiber modulus

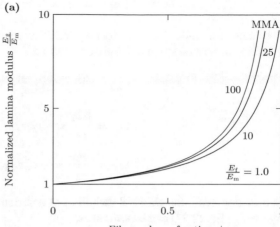

(a)

$\frac{E_2}{E_m}$ Normalized lamina modulus

MMA

25

100

10

$\frac{E_f}{E_m} = 1.0$

Fiber volume fraction ϕ_f

(b)

$\frac{E_2}{E_f}$ Normalized lamina modulus

MMA

$\frac{E_f}{E_m} = 1.0$

10

25

100

Fiber volume fraction ϕ_f

Fig. 2.7 Mechanical models to calculate the major Poisson's ratio

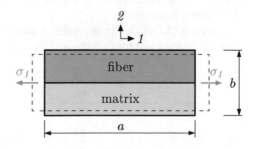

fiber

matrix

σ_1 σ_1

a

b

2

1

$$\Delta b = \Delta b_f + \Delta b_m \ (< 0 \text{ for } \sigma_1 > 0) \,. \tag{2.22}$$

The strains perpendicular to the loading direction can be stated based on the general definition of normal strain and Poisson's ratio as:

$$\varepsilon_2 = \frac{\Delta b}{b} = -\nu_{12}\varepsilon_1 \,, \tag{2.23}$$

$$\varepsilon_{f,2} = \frac{\Delta b_f}{b_f} = -\nu_f \varepsilon_{f,1} \,, \tag{2.24}$$

$$\varepsilon_{m,2} = \frac{\Delta b_m}{b_m} = -\nu_m \varepsilon_{m,1} \,. \tag{2.25}$$

Considering again the classical definition of a normal strain and that $\phi_f b = b_f$ and $\phi_m b = b_m$, Eq. (2.22) can be written as:

$$\Delta b = \varepsilon_{f,2} b_f + \varepsilon_{m,2} \tag{2.26}$$

$$= \varepsilon_{f,2}\phi_f b + \varepsilon_{m,2}\phi_m b \,, \tag{2.27}$$

which results after division by b and the consideration of the definitions in Eqs. (2.24)-(2.25) to the following expression:

$$\frac{\Delta b}{b} = \varepsilon_2 = -\nu_f \underbrace{\varepsilon_{f,1}}_{\varepsilon_1} \phi_f - \nu_m \underbrace{\varepsilon_{m,1}}_{\varepsilon_1} \phi_m \,, \tag{2.28}$$

which gives finally under consideration of $\varepsilon_2/\varepsilon_1 = -\nu_{12}$:

$$\nu_{12} = \nu_f \phi_f + \nu_m (1 - \phi_f) \,. \tag{2.29}$$

A graphical representation of the estimation for the major (lamina) Poisson's ratio is given in Fig. 2.8. A linear transition between the ratio of the matrix and the ratio of the fibers is obtained.

Equation (2.29) can be normalized either by the ratio of the matrix or the ratio of the fibers to obtain the following normalized representations of the Poisson's ratio:

$$\frac{\nu_{12}}{\nu_m} = \frac{\nu_f}{\nu_m}\phi_f + (1 - \phi_f) \,, \tag{2.30}$$

$$\frac{\nu_{12}}{\nu_f} = \phi_f + \frac{\nu_m}{\nu_f}(1 - \phi_f) \,, \tag{2.31}$$

which allows a graphical representation of the major Poisson's ratio for different proportions of the ratios, see Fig. 2.9.

Fig. 2.8 Major (lamina)
Poisson's ratio as a function
of the fiber volume fraction
(MMA)

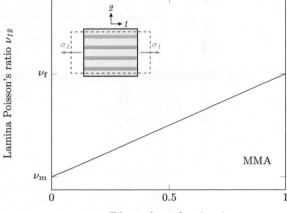

2.1.4   In-Plane Shear Modulus

To estimate the in-plane shear modulus (G_{12}), it is assumed for simplicity that the
shear stress of the fiber $(\tau_{f,12})$, matrix $(\tau_{m,12})$ as well as the entire composite (τ_{12}) are
the same over the entire cross section,[2] see Fig. 2.10 for schematic representations
of the constituents.

The shear stresses, which are based on the assumption to be equal $(\tau_{f,12} = \tau_{m,12} = \tau_{12})$, can be expressed due to Hooke's law for a pure shear state as

$$\tau_{12} = G_{12}\gamma_{12}, \tag{2.32}$$

$$\tau_{f,12} = G_{f}\gamma_{f,12}, \tag{2.33}$$

$$\tau_{m,12} = G_{m}\gamma_{m,12}, \tag{2.34}$$

whereas the shear strains can be approximated as

$$\gamma_{12} \approx \frac{\Delta a}{b}, \tag{2.35}$$

$$\gamma_{f,12} \approx \frac{\Delta a_{f}}{b_{f}}, \tag{2.36}$$

$$\gamma_{m,12} \approx \frac{\Delta a_{m}}{b_{m}}. \tag{2.37}$$

---

[2] This simplification is similar to the Timoshenko beam theory, which assumes a constant shear
stress over the beam, see [5].

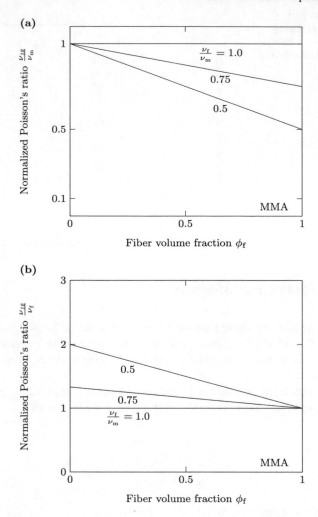

Fig. 2.9 Normalized major (lamina) Poisson's ratio as a function of the fiber volume fraction (MMA): **a** normalized by matrix ratio, **b** normalized by fiber ratio

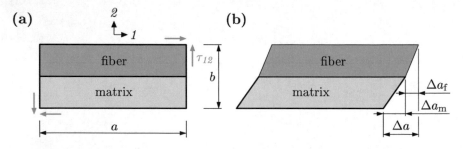

Fig. 2.10 Mechanical models to calculate the in-plane shear modulus: **a** undeformed geometry and load condition, **b** deformed shape

Fig. 2.11 In-plane shear modulus as a function of the fiber volume fraction (MMA)

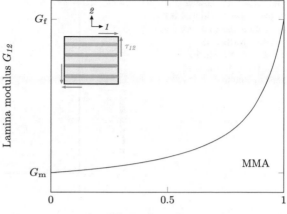

The total horizontal deformation can be expressed as

$$\Delta a = \Delta a_{\mathrm{f}} + \Delta a_{\mathrm{m}}, \tag{2.38}$$

or after dividing by b as:

$$\frac{\Delta a}{b} = \frac{\Delta a_{\mathrm{f}}}{b} + \frac{\Delta a_{\mathrm{m}}}{b}. \tag{2.39}$$

Considering the definitions of the normal strains (see Eqs. (2.35)–(2.37)) and Hooke's law (see Eqs. (2.32)–(2.34)), as well as that $\phi_{\mathrm{f}}b = b_{\mathrm{f}}$ and $\phi_{\mathrm{m}}b = b_{\mathrm{m}}$, Eq. (2.39) can be written as:

$$\gamma_{12} = \frac{\tau_{12}}{G_{12}} = \gamma_{\mathrm{f},12}\frac{b_{\mathrm{f}}}{b} + \gamma_{\mathrm{m},12}\frac{b_{\mathrm{m}}}{b} \tag{2.40}$$

$$= \frac{\tau_{\mathrm{f},12}}{G_{\mathrm{f}}}\phi_{\mathrm{f}} + \frac{\tau_{\mathrm{m},12}}{G_{\mathrm{m}}}\phi_{\mathrm{m}}, \tag{2.41}$$

or finally as:

$$G_{12} = \frac{G_{\mathrm{f}}G_{\mathrm{m}}}{G_{\mathrm{m}}\phi_{\mathrm{f}} + G_{\mathrm{f}}(1 - \phi_{\mathrm{f}})}. \tag{2.42}$$

A graphical representation of the estimation for the in-plane shear modulus is given in Fig. 2.11. A nonlinear transition between the modulus of the matrix and the modulus of the fibers is obtained.

Fig. 2.12 Normalized in-plane shear modulus as a function of the fiber volume fraction (MMA): **a** normalized by matrix modulus, **b** normalized by fiber modulus

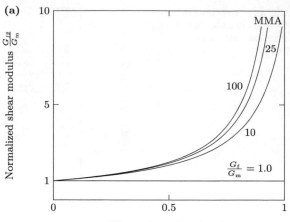

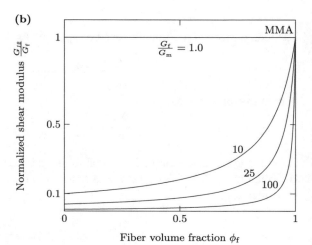

Equation (2.42) can be normalized either by the modulus of the matrix or the modulus of the fibers to obtain the following normalized representations of the in-plane shear modulus:

$$\frac{G_{12}}{G_{\mathrm{m}}} = \frac{1}{\frac{G_{\mathrm{m}}}{G_{\mathrm{f}}}\phi_{\mathrm{f}} + (1 - \phi_{\mathrm{f}})}, \tag{2.43}$$

$$\frac{G_{12}}{G_{\mathrm{f}}} = \frac{1}{\phi_{\mathrm{f}} + \frac{G_{\mathrm{f}}}{G_{\mathrm{m}}}(1 - \phi_{\mathrm{f}})}, \tag{2.44}$$

which allows a graphical representation of the in-plane shear modulus for different ratios of the moduli, see Fig. 2.12.

2.2 Elasticity Solutions with Contiguity After Tsai

The theory of elasticity solutions with contiguity after Tsai et al. [8] considers that the fibers do not remain perfectly straight and parallel during the manufacturing process of the composite. In addition, fibers are in contact with one another and not perfectly surrounded and isolated by the matrix. These geometric deviations are considered by two factor, i.e., the empirical factor k, which accounts for the fiber misalignment, and the factor C, which accounts for the fiber contiguity, see Fig. 2.13.

2.2.1 Modulus of Elasticity in Fiber Direction

The prediction of the modulus of elasticity in fiber direction is based on the classical expression (2.8), which is now corrected by the fiber misalignment such that

$$E_l = k \times (E_f \phi_f + E_m(1 - \phi_f)) , \tag{2.45}$$

where k is the filament (fiber) misalignment factor. Typically, this factor, which considers that the fibers are not perfectly parallel or not straight, is in the range of $0.9 \leq k \leq 1$, see [4]. The graphical representation of this modulus is given in Fig. 2.14 for different values of the fiber misalignment factor k. It should be noted here that the curve for $k = 1$ is identical with the representation in Fig. 2.2.

2.2.2 Modulus of Elasticity Transverse to Fiber Direction

The derivation considers two different limiting cases which are finally linear superposed. The first contribution to the modulus of elasticity transverse to fiber direction comes from the assumption that the fibers are isolated and the matrix is contiguous.

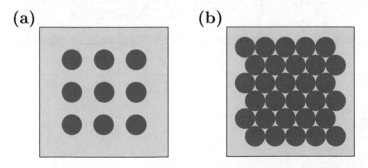

Fig. 2.13 **a** Perfectly isolated fibers ($C = 0$) and **b** contiguous fibers ($C = 1$). Adapted from [4]

Fig. 2.14 Modulus of
elasticity in fiber direction as
a function of the fiber
volume fraction (Tsai)

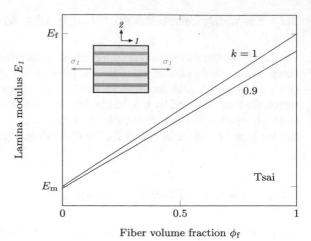

Fiber volume fraction ϕ_f

This means that $C = 0$ (see Fig. 2.13a) and the corresponding lower bound for E_2 can be obtained as follows:

$$E_2^{(m)} = 2[1 - \nu_2] \times \frac{K_f(2K_m + G_m) - G_m(K_f - K_m)\phi_m}{(2K_m + G_m) + 2(K_f - K_m)\phi_m}, \qquad (2.46)$$

where $\nu_2 = \nu_f - (\nu_f - \nu_m)\phi_m$ is Poisson's ratio in the isotropic plane normal to the fibers and this value should not be confused with the major Poisson's ratio (ν_{12}) of the unidirectional lamina.

The upper bound for E_2 can be obtained by the assumption $C = 1$ (see Fig. 2.13b) and is obtained from Eq. (2.46) by interchanging the subscripts 'm' and 'f' and replacing ϕ_m by $(1 - \phi_m)$:

$$E_2^{(f)} = 2[1 - \nu_2] \times \frac{K_m(2K_f + G_f) - G_f(K_m - K_f)(1 - \phi_m)}{(2K_f + G_f) + 2(K_m - K_f)(1 - \phi_m)}$$

$$= 2[1 - \nu_2] \times \frac{K_f(2K_m + G_f) + G_f(K_m - K_f)\phi_m}{(2K_m + G_f) - 2(K_m - K_f)\phi_m}. \qquad (2.47)$$

Linear superposition of expressions (2.46) and (2.46) gives the final expresion for E_2 as

$$E_2 = E_2^{(m)} + C(E_2^{(f)} - E_2^{(m)}) = (1 - C)E_2^{(m)} + CE_2^{(f)}, \qquad (2.48)$$

or in components:

Fig. 2.15 Modulus of
elasticity transverse to fiber
direction as a function of the
fiber volume fraction (Tsai)

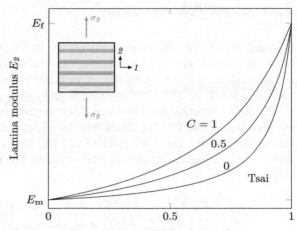

$$E_2 = 2\left[1 - \nu_f + (\nu_f - \nu_m)\phi_m\right] \times$$

$$\times \left[(1-C)\frac{K_f(2K_m + G_m) - G_m(K_f - K_m)\phi_m}{(2K_m + G_m) + 2(K_f - K_m)\phi_m} + \right. \quad (2.49)$$

$$\left. + C\frac{K_f(2K_m + G_f) + G_f(K_m - K_f)\phi_m}{(2K_m + G_f) - 2(K_m - K_f)\phi_m}\right],$$

where the elastic moduli can be calculated according to [4, 8] as

$$K_f = \frac{E_f}{2(1 - \nu_f)}, \quad (2.50)$$

$$G_f = \frac{E_f}{2(1 + \nu_f)}, \quad (2.51)$$

$$K_m = \frac{E_m}{2(1 - \nu_m)}, \quad (2.52)$$

$$G_m = \frac{E_m}{2(1 + \nu_m)}, \quad (2.53)$$

and the fiber contiguity is in the range of $0 \leq C \leq 1$. The graphical representation
of the modulus E_2 is given in Fig. 2.15 for different values of the fiber contiguity C.

2.2.3   Major Poisson's Ratio

It is important to take the relationship for the major Poisson's ratio from Ref. [7] since the classical textbook by Jones [4] contains a few typos[3] for this relation. In addition, scientific literature contains more faulty versions for ν_{12} which will be not mentioned here.

The derivation considers again two different limiting cases which are finally linear superposed. The first contribution to the major Poisson's ratio comes from the assumption that the fibers are isolated and the matrix is contiguous. This means that $C = 0$ (see Fig. 2.13a) and the corresponding bound for ν_{12} can be obtained as follows:

$$\nu_{12}^{(m)} = \frac{K_{\mathrm{f}}\nu_{\mathrm{f}}(2K_{\mathrm{m}} + G_{\mathrm{m}})(1 - \phi_{\mathrm{m}}) + K_{\mathrm{m}}\nu_{\mathrm{m}}(2K_{\mathrm{f}} + G_{\mathrm{m}})\phi_{\mathrm{m}}}{K_{\mathrm{m}}(2K_{\mathrm{f}} + G_{\mathrm{m}}) + G_{\mathrm{m}}(K_{\mathrm{f}} - K_{\mathrm{m}})(1 - \phi_{\mathrm{m}})}. \tag{2.54}$$

The other bound for ν_{12} can be obtained by the assumption $C = 1$ (see Fig. 2.13b) and is obtained from Eq. (2.54) by interchanging the subscripts 'm' and 'f' and replacing ϕ_{m} by $(1 - \phi_{\mathrm{m}})$:

$$\nu_{12}^{(f)} = \frac{K_{\mathrm{m}}\nu_{\mathrm{m}}(2K_{\mathrm{f}} + G_{\mathrm{f}})\phi_{\mathrm{m}} + K_{\mathrm{f}}\nu_{\mathrm{f}}(2K_{\mathrm{m}} + G_{\mathrm{f}})(1 - \phi_{\mathrm{m}})}{K_{\mathrm{f}}(2K_{\mathrm{m}} + G_{\mathrm{f}}) + G_{\mathrm{f}}(K_{\mathrm{m}} - K_{\mathrm{f}})\phi_{\mathrm{m}}}. \tag{2.55}$$

Linear superposition of expressions (2.54) and (2.55) gives the final expresion for ν_{12} as

$$\nu_{12} = \nu_{12}^{(m)} + C(\nu_{12}^{(f)} - \nu_{12}^{(m)}) = (1 - C)\nu_{12}^{(m)} + C\nu_{12}^{(f)}, \tag{2.56}$$

or in components:

$$\nu_{12} = (1 - C)\frac{K_{\mathrm{f}}\nu_{\mathrm{f}}(2K_{\mathrm{m}} + G_{\mathrm{m}})\phi_{\mathrm{f}} + K_{\mathrm{m}}\nu_{\mathrm{m}}(2K_{\mathrm{f}} + G_{\mathrm{m}})\phi_{\mathrm{m}}}{K_{\mathrm{m}}(2K_{\mathrm{f}} + G_{\mathrm{m}}) + G_{\mathrm{m}}(K_{\mathrm{f}} - K_{\mathrm{m}})\phi_{\mathrm{f}}} +$$
$$+ C\frac{K_{\mathrm{m}}\nu_{\mathrm{m}}(2K_{\mathrm{f}} + G_{\mathrm{f}})\phi_{\mathrm{m}} + K_{\mathrm{f}}\nu_{\mathrm{f}}(2K_{\mathrm{m}} + G_{\mathrm{f}})\phi_{\mathrm{f}}}{K_{\mathrm{f}}(2K_{\mathrm{m}} + G_{\mathrm{f}}) + G_{\mathrm{f}}(K_{\mathrm{m}} - K_{\mathrm{f}})\phi_{\mathrm{m}}}. \tag{2.57}$$

The graphical representation of the major (lamina) Poisson's ratio ν_{12} is given in Fig. 2.16 for different values of the fiber contiguity C.

---

[3] The faulty equation reads:

$$\nu_{12} = (1 - C)\frac{K_{\mathrm{f}}\nu_{\mathrm{f}}(2K_{\mathrm{m}} + G_{\mathrm{m}})\phi_{\mathrm{f}} + K_{\mathrm{m}}\nu_{\mathrm{m}}(2K_{\mathrm{f}} + G_{\mathrm{m}})\phi_{\mathrm{m}}}{K_{\mathrm{f}}(2K_{\mathrm{m}} + G_{\mathrm{m}}) - G_{\mathrm{m}}(K_{\mathrm{f}} - K_{\mathrm{m}})\phi_{\mathrm{m}}} +$$
$$+ C\frac{K_{\mathrm{m}}\nu_{\mathrm{m}}(2K_{\mathrm{f}} + G_{\mathrm{f}})\phi_{\mathrm{m}} + K_{\mathrm{f}}\nu_{\mathrm{f}}(2K_{\mathrm{m}} + G_{\mathrm{f}})\phi_{\mathrm{f}}}{K_{\mathrm{f}}(2K_{\mathrm{m}} + G_{\mathrm{m}}) + G_{\mathrm{f}}(K_{\mathrm{m}} - K_{\mathrm{f}})\phi_{\mathrm{m}}}.$$

Fig. 2.16 Major (lamina) Poisson's ratio as a function of the fiber volume fraction (Tsai)

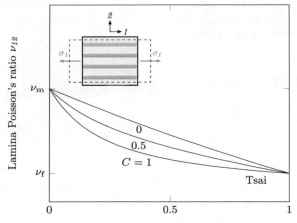

2.2.4 In-Plane Shear Modulus

The original reference by Tsai [7] states only an upper and lower bound for the in-plane shear modulus, i.e.,

$$G^{\text{upper}} = \left(\frac{1}{E_1} + \frac{2\nu_{12}}{E_1} + \frac{1}{E_2} \right)^{-1} \Bigg|_{C=k=1} , \tag{2.58}$$

$$G^{\text{lower}} = G_{\text{m}} + \frac{15(1 - \nu_{\text{m}}) \left(\dfrac{G_{\text{f}}}{G_{\text{m}}} - 1 \right)(1 - \phi_{\text{m}})G_{\text{m}}}{7 - 5\nu_{\text{m}} + 2(4 - 5\nu_{\text{m}}) \left[1 + \left(\dfrac{G_{\text{f}}}{G_{\text{m}}} - 1 \right)\phi_{\text{m}} \right]}. \tag{2.59}$$

Thus, the follwoing provides the final prediction which is again derived by considering two limiting cases, i.e., $C = 0$ and $C = 1$, and the subsequent superposition [8]:

$$\begin{aligned} G_{12} = (1 - C)G_{\text{m}} & \frac{2G_{\text{f}} - (G_{\text{f}} - G_{\text{m}})\phi_{\text{m}}}{2G_{\text{m}} + (G_{\text{f}} - G_{\text{m}})\phi_{\text{m}}} + \\ & + CG_{\text{f}}\frac{(G_{\text{f}} + G_{\text{m}}) - (G_{\text{f}} - G_{\text{m}})\phi_{\text{m}}}{(G_{\text{f}} + G_{\text{m}}) + (G_{\text{f}} - G_{\text{m}})\phi_{\text{m}}}. \end{aligned} \tag{2.60}$$

The graphical representation of the in-plane shear modulus G_{12} is given in Fig. 2.17 for different values of the fiber contiguity C.

Fig. 2.17 In-plane shear
modulus as a function of the
fiber volume fraction (Tsai)

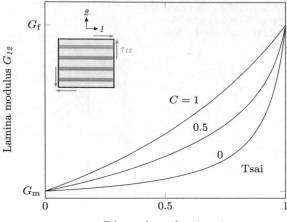

2.3 Halpin-Tsai Relationships

The so-called Halpin-Tsai approximations [1, 4] are based on a mathematical model
by Hermans [2], which consists of concentric cylinders surrounded by an unbounded
composite. Based on an interpolation procedure, reasonable accurate approximations
of the original predictions by Hermans were obtained, which can be applied over a
wide range of fiber volume fractions.

2.3.1 Modulus of Elasticity in Fiber Direction and Major
Poisson's Ratio

The Halpin-Tsai relationships for the modulus of elasticity in fiber direction and the
major Poisson's ratio are identical to the expressions obtained by the mechanics of
materials approach, see Eqs. (2.8) and (2.29) and Figs. 2.2 and 2.8:

$$E_1 = E_f \phi_f + E_m (1 - \phi_f) \,, \tag{2.61}$$
$$\nu_{12} = \nu_f \phi_f + \nu_m (1 - \phi_f) \,. \tag{2.62}$$

2.3.2 *Modulus of Elasticity Transverse to Fiber Direction and In-Plane Shear Modulus*

The Halpin-Tsai relationships for the modulus of elasticity transverse to the fiber direction and the in-plane shear modulus have the same structure, i.e.,

$$E_2 = E_\mathrm{m} \times \frac{1 + \xi \eta \phi_\mathrm{f}}{1 - \eta \phi_\mathrm{f}}, \tag{2.63}$$

$$G_{12} = G_\mathrm{m} \times \frac{1 + \xi \eta \phi_\mathrm{f}}{1 - \eta \phi_\mathrm{f}}, \tag{2.64}$$

where the coefficient η is given by

$$\eta = \frac{\frac{E_\mathrm{f}}{E_\mathrm{m}} - 1}{\frac{E_\mathrm{f}}{E_\mathrm{m}} + \xi}, \quad (\text{for } E_2) \tag{2.65}$$

$$\eta = \frac{\frac{G_\mathrm{f}}{G_\mathrm{m}} - 1}{\frac{G_\mathrm{f}}{G_\mathrm{m}} + \xi}. \quad (\text{for } G_{12}) \tag{2.66}$$

The parameter ξ is a measure of the reinforcement and depends on the boundary conditions, i.e., geometry of inclusions and loading conditions. The graphical representations of the moduli E_2 and G_{12} are given in Figs. 2.18 and 2.19 for different values of the parameter ξ.

Fig. 2.18 Modulus of elasticity transverse to fiber direction as a function of the fiber volume fraction (Halpin-Tsai)

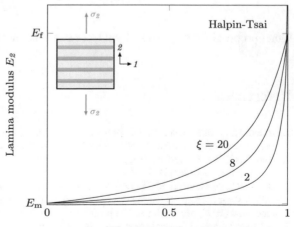

Fig. 2.19 In-plane shear
modulus as a function of the
fiber volume fraction
(Halpin-Tsai)

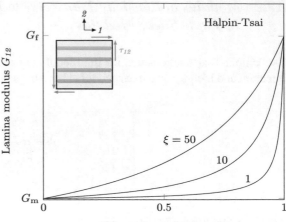

There are different predictions and estimates of factor ξ, which in general may vary between $\xi = 0$ and $\xi = \infty$, see [1]. For circular fibers in a square array, the values $\xi_{E_2} = 2$ for the modulus E_2 and $\xi_{G_{12}} = 1$ for the modulus G_{12} were obtained to provide reasonable results. In the case of rectangular cross section fibers of aspect ration a/b (width a and thickness b) in a hexagonal array, the following estimates provide good approximations:

$$\xi_{E_2} = 2 \times \frac{a}{b}, \tag{2.67}$$

$$\log \left(\xi_{G_{12}} \right) = \sqrt{3} \times \log \left(\frac{a}{b} \right). \tag{2.68}$$

Other proposal for $\xi_{G_{12}}$ can be found, for example, in [3].

References

1. Halpin JC (1969) Effects of environmental factors on composite materials. Air Force Materials Laboratory Technical Report AFML-TR-67-423. Air Force Materials Laboratory, Ohio
2. Hermans JJ (1967) The elastic properties of fiber reinforced materials when the fibers are aligned. P K Akad Wet-Amsterd B70:1–9
3. Hewitt RL, de Malherbe MC (1970) A review and catalogue of an approximation for the longitudinal shear modulus of continuous fibre composites. J Compos Mater 4:280–282
4. Jones RM (1999) Mechanics of composite materials. Taylor & Francis, Philadelphia
5. Öchsner A (2021) Classical beam theories of structural mechanics. Springer, Cham
6. Paul B (1960) Prediction of elastic constants of multiphase materials. T Metall Soc AIME 218:36–41

7. Tsai SW, Springer GS, Schultz AB (1963) The composite behavior of filament-wound materials. In: Fourteenth international astronautical congress, Paris, France, September 1963, Paper No. 139
8. Tsai SW (1964) Structural behavior of composite materials. Technical report NASA-CR-71, Washington

Chapter 3
Comparison with Experimental Results

Abstract This chapter presents two classical experimental data sets for the modulus of elasticity in and transverse to the fiber direction, the major Poisson's ratio, as wells as the in-plane shear modulus. These data sets were reported by Tsai and Whitney/Riley. The experimental values are used to benchmark the theoretical predications based on the mechanics of materials approach, the elasticity solutions with contiguity after Tsai, and the Halpin-Tsai relationships. The chapter concludes with the presentation of optimized material parameters (i.e., fitting parameters) to predict the corresponding properties as a function of the fiber volume fraction.

3.1 Extraction of Experimental Values

The following sections report two classical data sets for the elastic lamina properties as function of some fraction. These data sets go back to Tsai [2, 3] and Whitney and Riley [4]. The tables in the following subsections report the experimental data as given in the corresponding publications and in addition, the conversion to SI units and other reference numbers. Since some experimental data was reported only in the form of diagrams,[1] it is hoped that the collection of numerical values in tables allows an easier further processing of the experimental results.

3.1.1 Data Set by Tsai

The data set by Tsai [2, 3] was obtained from unidirectional glass filamant-epoxy resin systems, which were manufactured by the hand lay-up technique. Scotch-ply No. 1009-33 W2 38 (Minnesota Mining and Manufacturing Company, U.S.A.) in the range $0.2 \leq \psi_m \leq 0.35$ and E-787-NUF (U.S. Polymeric Company) in the range $0.13 \leq \psi_m \leq 0.20$ were used. The properties of the constituents, i.e., fiber and matrix, are summarized in Table 3.1.

---

[1] These diagrams were digitalized and a specialized software (Digitizeit) was used to extract the numerical values of data points.

© The Author(s), under exclusive license to Springer Nature Switzerland AG 2022   29
A. Öchsner, *Micromechanics of Fiber-Reinforced Laminae*,
SpringerBriefs in Continuum Mechanics,
https://doi.org/10.1007/978-3-030-94091-1_3

Table 3.1 Fiber and matrix properties of a glass filament-epoxy resin system, [2, 3]

	Fiber	Matrix
E in 10^6 psi	10.6	0.5
in MPa	73084.42	3447.38
v in –	0.22	0.35
ρ in –	2.6	1.15

Table 3.2 Experimental results for the lamina modulus E_1 in case of a glass filament-epoxy resin system, [2, 3]

ψ_{m} in % [2, 3]	E_1 in 10^6 psi [2, 3]	ϕ_{f}	E_1 in MPa
12.9	7.53	0.749	51900
13.8	8.02	0.735	55300
14.9	7.58	0.716	52200
14.9	7.14	0.716	49300
14.9	6.83	0.716	47100
16.0	6.98	0.699	48200
16.1	6.60	0.698	45500
16.8	7.63	0.687	52600
19.9	6.65	0.640	45900
20.9	5.88	0.626	40500
21.9	6.11	0.612	42100
21.9	6.84	0.611	47200
23.0	6.56	0.596	45200
23.1	6.84	0.596	47200
23.2	6.19	0.594	42700
23.3	5.87	0.593	40400
28.8	5.32	0.522	36700

The values for the moduli E_1 (see Table 3.2) and E_2 (see Table 3.3) were obtained from bending or uniaxial tension tests on 0° or 90° specimens (machined parallel or transverse to the fibers). The strain measurement was done by strain rosettes or the cross-head movement.

The values for the lamina Poisson's ratio v_{12} (see Table 3.4) were obtained from bending or uniaxial tension tests on 0° specimens whereas the strain measurement was done by strain rosettes.

The values for the lamina in-plane shear modulus G_{12} (see Table 3.5) were obtained from pure twisting of 0° square plates. The experiments were realized by four corner forces which were perpendicular to the plate (two upwards on opposing corners and two downwards on opposing corners).

Table 3.3 Experimental results for the lamina modulus E_2 in case of a glass filament-epoxy resin system, [2, 3]

ψ_m in % [2, 3]	E_2 in 10^6 psi [2, 3]	ϕ_f	E_2 in MPa
12.5	2.87	0.755	19800
14.4	2.58	0.724	17800
15.0	2.29	0.715	15800
15.1	2.77	0.713	19100
15.1	1.98	0.713	13700
15.1	3.10	0.713	21400
16.0	2.12	0.700	14600
16.0	1.86	0.699	12800
16.1	2.53	0.698	17400
17.2	2.55	0.680	17600
18.2	2.50	0.665	17200
20.0	1.85	0.638	12700
20.1	2.13	0.638	14700
21.1	1.61	0.623	11100
22.0	2.40	0.611	16500
22.1	1.81	0.609	12500
23.0	2.04	0.596	14000
23.1	2.36	0.595	16300

Table 3.4 Experimental results for the lamina Poisson's ratio ν_{12} in case of a glass filament-epoxy resin system, [2, 3]

ψ_m in % [2, 3]	ν_{12} [2, 3]	ϕ_f
13.8	0.247	0.734
14.7	0.243	0.719
15.7	0.244	0.704
16.2	0.251	0.695
22.6	0.250	0.602
30.7	0.277	0.500
33.2	0.284	0.471

3.1.2 Data Set by Whitney and Riley

The data set by Whitney and Riley [4] was obtained from unidirectional boron filament-epoxy resin systems (Union Carbide ERL 2256-ZZLB 0820), which were manufactured either by the hand lay-up technique or wound on a special machine. The properties of the constituents, i.e., fiber and matrix, are summarized in Table 3.6.

Table 3.5 Experimental results for the lamina in-plane shear modulus G_{12} in case of a glass filament-epoxy resin system, [2, 3]

ψ_m in % [2, 3]	G_{12} in 10^6 psi [2, 3]	ϕ_f	G_{12} in MPa
14.1	1.49	0.729	10300
13.8	1.24	0.734	8580
14.5	1.28	0.723	8860
14.0	1.08	0.730	7410
16.0	1.22	0.698	8410
17.7	1.25	0.672	8600
23.2	1.03	0.595	7110
23.1	0.945	0.595	6520

Table 3.6 Fiber and matrix properties of a boron filament-epoxy resin system, [4]

	Fiber	Matrix
E in 10^6 psi	60.0	0.6
in MPa	413685.42	4136.85
v in –	0.20	0.35
G in 10^6 psi	25.0	0.22
in MPa	172368.93	1516.85

Table 3.7 Experimental results for the lamina properties in case of a boron filament-epoxy resin system, as reported in [4]

ϕ_f in %	E_1 in 10^6 psi	E_2 in 10^6 psi	G_{12} in 10^6 psi	v_{12} in –
0.20	11.7	–	–	0.154
0.55	30.1	–	–	0.141
0.60	35.7	3.10	–	0.133
0.65	35.5	3.40	–	0.130
0.70	34.5	3.88	1.77	0.125
0.75	–	4.90	2.43	–

The values for the modulus E_1 and for the lamina Poisson's ratio v_{12} (see Table 3.7 for the original data and Table 3.8 for the data in SI units) were obtained from single layer tensile specimens (manufactured by the hand lay-up technique for high fiber volume content and winding for low fiber content). The strain measurement was done by strain gages. The values for the transverse modulus E_2 were obtained from wound specimens of 3 to 8 layers. The values for the lamina in-plane shear modulus G_{12} were calculated in the following way: filament wound specimens were cut in such a way that the fiber direction was at 45° to the sides. These particular specimens were used to measure the tensile modulus at 45° to the fibers. Based on these results, the shear modulus G_{12} was calculated.

Table 3.8 Experimental results for the lamina properties in case of a boron filament-epoxy resin system (converted to SI units), [4]

ϕ_f in %	E_1 in MPa	E_2 in MPa	G_{12} in MPa	ν_{12} in –
0.20	80669	–	–	0.154
0.55	207532	–	–	0.141
0.60	246143	21374	–	0.133
0.65	244764	23442	–	0.130
0.70	237869	26752	12204	0.125
0.75	–	33784	16754	–

3.2 Comparison Between Theoretical Predictions and Experimental Results

3.2.1 Data Set by Tsai

The approximation of the modulus of elasticity in fiber direction (E_1) as a function of the fiber volume fraction is presented in Fig. 3.1 based on the predictions of Eqs. (2.8), (2.45), and (2.61). It can be seen that the simple straight line approximation between E_m and E_f gives already a quite good representation of the experimental data points by Tsai. The fiber misalignment factor k allows some kind of fine tuning of the prediction, see Fig. 3.1b.

The approximation of the modulus of elasticity transverse to the fiber direction (E_2) as a function of the fiber volume fraction is presented in Fig. 3.2 based on the predictions of Eqs. (2.18), (2.49), and (2.63). It an be seen that the prediction based on the mechanics of materials approach (see Fig. 3.2a) clearly underestimates the experimental data. The fiber contiguity (C) or the Halpin-Tsai coefficient (ξ) allow a much better adjustment to the experimental points, see Fig. 3.2b, c.

The approximation of the Lamina Poisson's ratio (ν_{12}) as a function of the fiber volume fraction is presented in Fig. 3.3 based on the predictions of Eqs. (2.29), (2.57), and (2.62). It can be seen that the simple straight line approximation between ν_m and ν_f gives already a quite good representation of the experimental data points by Tsai, possibly with a small tendency to overestimate the values, see Fig. 3.3a, c. The fiber contiguity C allows some kind of fine tuning of the prediction, see Fig. 3.3b.

The approximation of the in-plane shear modulus (G_{12}) as a function of the fiber volume fraction is presented in Fig. 3.4 based on the predictions of Eqs. (2.42), (2.60), and (2.64). It an be seen that the prediction based on the mechanics of materials approach (see Fig. 3.4a) clearly underestimates the experimental data. The fiber contiguity (C) or the Halpin-Tsai coefficient (ξ) allow a much better adjustment to the experimental points, see Fig. 3.4b, c.

Fig. 3.1 Modulus of
elasticity in fiber direction as
a function of the fiber
volume fraction: comparison
of experimental values and
theoretical predictions (**a**
MMA, **b** Tsai, **c** Halpin-Tsai)

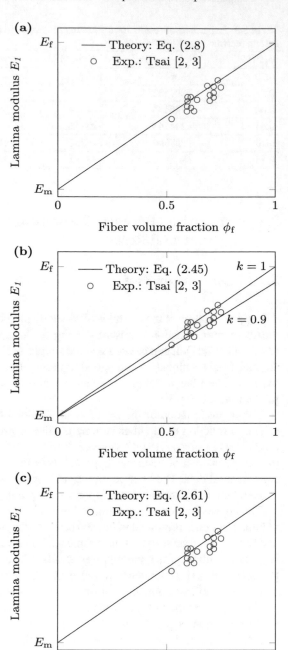

Fig. 3.2 Modulus of elasticity transverse to fiber direction as a function of the fiber volume fraction: comparison of experimental values and theoretical predictions (**a** MMΛ, **b** Tsai, **c** Halpin-Tsai)

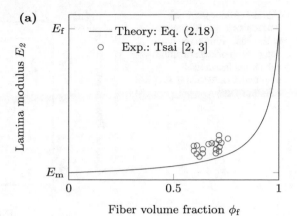

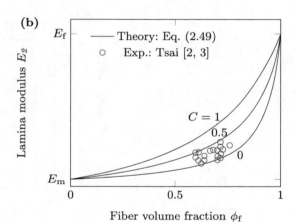

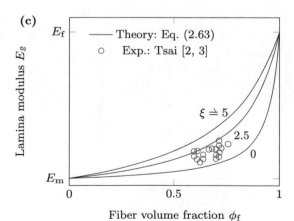

Fig. 3.3 Major (lamina)
Poisson's ratio as a function
of the fiber volume fraction:
comparison of experimental
values and theoretical
predictions (**a** MMA, **b** Tsai,
c Halpin-Tsai)

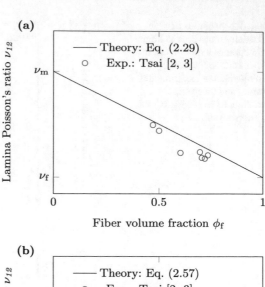

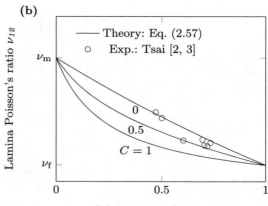

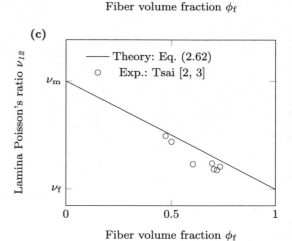

Fig. 3.4 In-plane shear
modulus as a function of the
fiber volume fraction:
comparison of experimental
values and theoretical
predictions (**a** MMA, **b** Tsai,
c Halpin-Tsai)

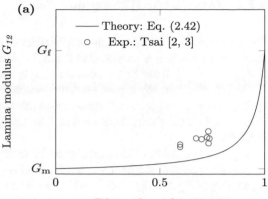

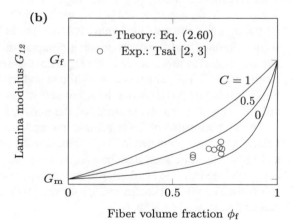

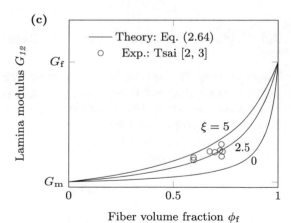

3.2.2 Data Set by Whitney and Riley

The approximation of the modulus of elasticity in fiber direction as a function of the fiber volume fraction is presented in Fig. 3.5 based on the predictions of Eqs. (2.8), (2.45), and (2.61). It can be seen that the simple straight line approximation between E_m and E_f gives already a quite good representation of the experimental data points by Tsai, possibly with a small tendency to overestimate the values. The fiber misalignment factor k allows some kind of fine tuning for $k < 1$ of the prediction, see Fig. 3.5b.

The approximation of the modulus of elasticity transverse to the fiber direction (E_2) as a function of the fiber volume fraction is presented in Fig. 3.6 based on the predictions of Eqs. (2.18), (2.49), and (2.63). It an be seen that the prediction based on the mechanics of materials approach (see Fig. 3.6a) underestimates the experimental data. The fiber contiguity (C) or the Halpin-Tsai coefficient (ξ) allow a much better adjustment to the experimental points, see Fig. 3.6b, c.

The approximation of the lamina Poisson's ratio (ν_{12}) as a function of the fiber volume fraction and the corresponding comparison with the experimental data is not meaningful in this case since the experimental results for Poisson's ratio are, as clearly stated in the publication by Whitney and Riley [4], obviously incorrect, see Fig. 3.7. It is not possible that the measured values are considerably less than the ratio of either constituent material, i.e., the fiber or the matrix.

The approximation of the in-plane shear modulus (G_{12}) as a function of the fiber volume fraction is presented in Fig. 3.8 based on the predictions of Eqs. (2.42), (2.60), and (2.64). It an be seen that the prediction based on the mechanics of materials approach (see Fig. 3.8a) underestimates the experimental data. The fiber contiguity (C) or the Halpin-Tsai coefficient (ξ) allow a much better adjustment to the experimental points, see Fig. 3.8b, c.

3.3 Optimized Representation of Theoretical Predictions

The expressions for the effective lamina moduli according to Eqs. (2.45), (2.49), (2.57), (2.60), (2.63), and (2.64) contain a single parameter, i.e., k, C, or ξ. This parameters can also be understood as a fitting parameter to approximate the proposed equations to a given data set from experiments. In a more mathematical sense, we can state that the task is to fit a set of data points $(x_i; y_i)$ $i = 1, \ldots, N$ to a linear or general model of the form

$$y(x) = y(x; a), \tag{3.1}$$

where y is one of the moduli mentioned in the above cited equations, x is the fiber volume fraction ($x = \phi_f$), and a is one of the fitting parameters, i.e., k, C, or ξ. To obtain the best value of the fitting parameter for a given data set, one may use, for

Fig. 3.5 Modulus of
elasticity in fiber direction as
a function of the fiber
volume fraction: comparison
of experimental values and
theoretical predictions (**a**
MMA, **b** Tsai, **c** Halpin-Tsai)

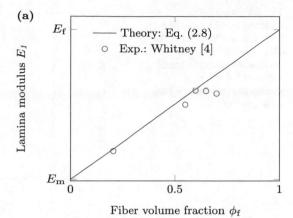

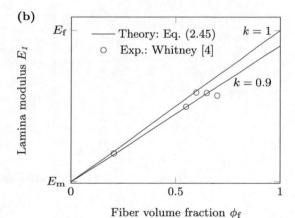

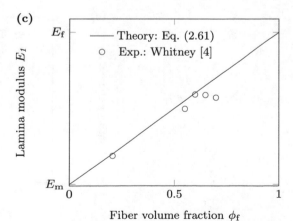

Fig. 3.6 Modulus of elasticity transverse to fiber direction as a function of the fiber volume fraction: comparison of experimental values and theoretical predictions (**a** MMA, **b** Tsai, **c** Halpin-Tsai)

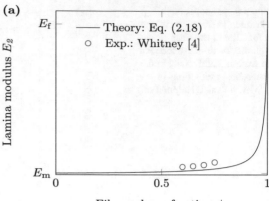

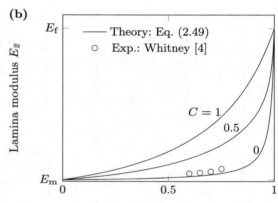

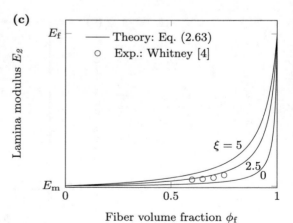

Fig. 3.7 Major (lamina)
Poisson's ratio as a function
of the fiber volume fraction:
comparison of experimental
values and theoretical
predictions (**a** MMA, **b** Tsai,
c Halpin-Tsai)

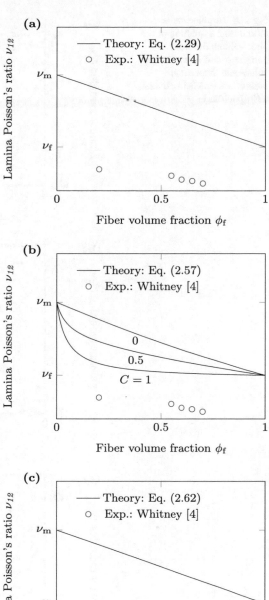

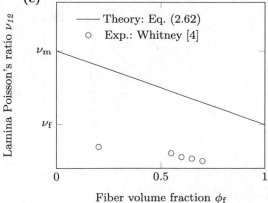

Fig. 3.8 In-plane shear modulus as a function of the fiber volume fraction: comparison of experimental values and theoretical predictions (**a** MMA, **b** Tsai, **c** Halpin-Tsai)

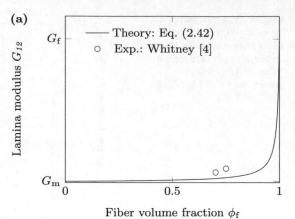

(a)

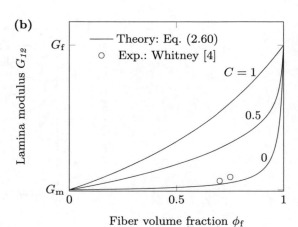

(b)

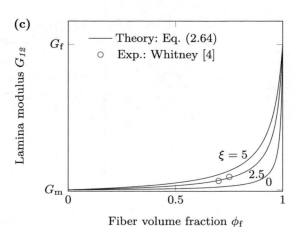

(c)

example, the method of least squares, which minimizes the sum of the squares of the errors, see [1, 5]:

$$\text{minimize} \sum_{i=1}^{N} [y_i - y(x_i; a)]^2 . \tag{3.2}$$

Many of the computer algebra systems such as Maple, Matlab® or Maxima, contain built-in functions to fit parameters in a regression analysis. This approach avoids a programming of the method of least squares.

3.3.1 Elasticity Solutions with Contiguity After Tsai

Considering the prediction of the modulus of elasticity in fiber direction according to Eq. (2.45) and the particular properties of fiber and matrix according to Tables 3.1 and 3.6, the following linear models are obtained:

$$E_1(\phi_f) = k \times \frac{3481852\phi_f + 172369}{50} \quad (\text{Exp.: Tsai}), \tag{3.3}$$

$$E_1(\phi_f) = k \times \frac{40954857\phi_f + 413685}{100} \quad (\text{Exp.: Whitney}), \tag{3.4}$$

where the parameter k (fiber misalignment factor) must be adjusted to the corresponding data sets. A graphical comparison between the experimental data sets and the theoretical predictions based on optimized k values is provided in Fig. 3.9.

Considering the prediction of the modulus of elasticity transverse to the fiber direction according to Eq. (2.49) and the particular properties of fiber and matrix according to Tables 3.1 and 3.6, the following cubic/quadratic rational functions are obtained:

$$E_2(\phi_f) = -\frac{(a_1 C + a_2)\phi_f^3 + (a_3 C + a_4)\phi_f^2 + (a_5 - a_6 C)\phi_f - a_7 C - a_8}{a_9\phi_f^2 - a_{10}\phi_f + a_{11}} \tag{3.5}$$
$$(\text{Exp.: Tsai}),$$

with

$$a_1 = 3135709377982879890894860017499640, \tag{3.6}$$
$$a_2 = 1396192485198600285230543944399456, \tag{3.7}$$
$$a_3 = 1254283751193152315971345672909462, \tag{3.8}$$
$$a_4 = 11259209763383325302217568138849624, \tag{3.9}$$

Fig. 3.9 Modulus of
elasticity in fiber direction as
a function of the fiber
volume fraction: comparison
of experimental values and
optimized theoretical
predictions according to Tsai
(**a** experiments by Tsai [2,
3], **b** experiments by
Whitney [4])

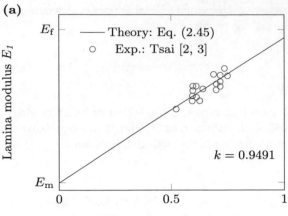

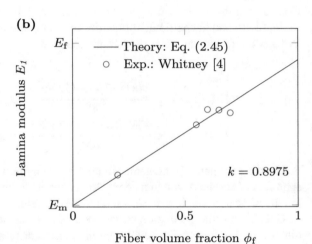

$$a_5 = 126745465575047842883128112819904619955\,,\tag{3.10}$$

$$a_6 = 156785468899143850787180776297055531\,,\tag{3.11}$$

$$a_7 = 18024569304188444205\,,\tag{3.12}$$

$$a_8 = 43583450647234175460057150948822755\,,\tag{3.13}$$

$$a_9 = 8411560061076986375602113796680\,,\tag{3.14}$$

$$a_{10} = 20804286127053137953373095674405\,,\tag{3.15}$$

$$a_{11} = 12642485205354249608988868936005\,,\tag{3.16}$$

as well as for the experimental data set by Whitney

$$E_2(\phi_f) = -\frac{(a_1 C + a_2)\,\phi_f^3 + (a_3 C + a_4)\,\phi_f^2 + (a_5 - a_6 C)\,\phi_f - a_7 C - a_8}{a_9 \phi_f^2 - a_{10}\phi_f + a_{11}}$$

$$(3.17)$$

(Exp.: Whitney),

with

$$a_1 = 99598716142554611958025682212\,, \tag{3.18}$$
$$a_2 = 893261113153788186901769504\,, \tag{3.19}$$
$$a_3 = 3319957204751823706411665142572\,, \tag{3.20}$$
$$a_4 = 6432778585281112596067173744\,, \tag{3.21}$$
$$a_5 = 60155315591632534584085178430\,, \tag{3.22}$$
$$a_6 = 43159443661773555255887073180066\,, \tag{3.23}$$
$$a_7 = 14478769212781406\,, \tag{3.24}$$
$$a_8 = 22040995730242621664780477672\,, \tag{3.25}$$
$$a_9 = 3886701242918683455377363\,, \tag{3.26}$$
$$a_{10} = 9193638060506547299070482\,, \tag{3.27}$$
$$a_{11} = 53279658992331499531538644\,. \tag{3.28}$$

Now the parameter C (fiber contiguity) must be adjusted to the corresponding data sets. A graphical comparison between the experimental data sets and the theoretical predictions based on optimized C values is provided in Fig. 3.10.

Considering the prediction of the major Poisson's ratio according to Eq. (2.57) and the particular properties of fiber and matrix according to Tables 3.1 and 3.6, the following quadratic/quadratic rational functions are obtained:

$$\nu_{12}(\phi_f) = \frac{(a_1 C - a_2)\,\phi_f^2 + (a_3 - a_4 C)\,\phi_f - a_5 C + a_6}{a_7 \phi_f^2 + a_8 \phi_f + a_9} \tag{3.29}$$

(Exp.: Tsai),

with

$$a_1 = 414060535731214663305394889372733 7966267\,, \tag{3.30}$$
$$a_2 = 272161701687253297429266777239366 2202875\,, \tag{3.31}$$
$$a_3 = 111286940428225327084688355587366 165345325\,, \tag{3.32}$$
$$a_4 = 414060535731211919232433509352664 1644301\,, \tag{3.33}$$
$$a_5 = 1585825920169636505705625 0\,, \tag{3.34}$$
$$a_6 = 292346875274519944268548599162708 8408150\,, \tag{3.35}$$

Fig. 3.10 Modulus of
elasticity transverse to fiber
direction as a function of the
fiber volume fraction:
comparison of experimental
values and *optimized*
theoretical predictions
according to Tsai (**a**
experiments by Tsai [2, 3], **b**
experiments by Whitney [4])

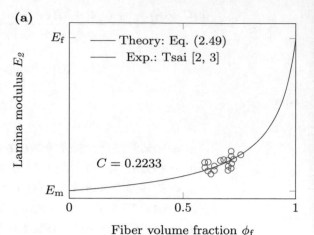

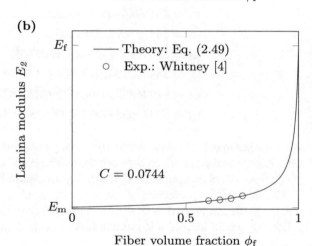

$$a_7 = 7555849134581552661327999526075070836700 \, , \tag{3.36}$$

$$a_8 = 3559386381268311566242008137580712918675 \, , \tag{3.37}$$

$$a_9 = 8352767864986289673215855054641582682425 \, , \tag{3.38}$$

as well as for the experimental data set by Whitney

$$\nu_{12}(\phi_{\mathrm{f}}) = \frac{(a_1 C - a_2)\, \phi_{\mathrm{f}}^2 + (a_3 - a_4 C)\, \phi_{\mathrm{f}} - a_5 C + a_6}{a_7 \phi_{\mathrm{f}}^2 + a_8 \phi_{\mathrm{f}} + a_9} \tag{3.39}$$

$$(\text{Exp.: Whitney}) \, ,$$

with

$$a_1 = 10961918217231003672758702880111 7122240 \,, \tag{3.40}$$
$$a_2 = 7586499307066753458818151286080091 4880 \,, \tag{3.41}$$
$$a_3 = 2550552917020068020210824776779258 61840 \,, \tag{3.42}$$
$$a_4 = 1096191821723095549834366063501899 32128 \,, \tag{3.43}$$
$$a_5 = 80166691910177419710 1518 \,, \tag{3.44}$$
$$a_6 = 1290150551671943043669745914966365 0640 \,, \tag{3.45}$$
$$a_7 = 1753251499854462655008817884531001 69920 \,, \tag{3.46}$$
$$a_8 = 7482724264213621970748932509769581 10604 \,, \tag{3.47}$$
$$a_9 = 368614443334841519861529716978501 38647 \,. \tag{3.48}$$

The parameter C (fiber contiguity) must be again adjusted to the corresponding data sets. A graphical comparison between the experimental data sets and the theoretical predictions based on optimized C values is provided in Fig. 3.11. Obviously, the experimental values by Whitney are not very meaningful since only values below the fiber ratio are reported. Nevertheless, some kind of reasonable fitting is obtained, which fulfills the boundary conditions for $\phi_f = 0$ and $\phi_f = 1$.

Considering the prediction of the in-plane shear modulus according to Eq. (2.60) and the particular properties of fiber and matrix according to Tables 3.1 and 3.6, the following quadratic/quadratic rational functions are obtained:

$$G_{12}(\phi_f) = -\frac{(a_1 C + a_2)\,\phi_f{}^2 + (-a_3 C - a_4)\,\phi_f - a_5}{a_6\phi_f{}^2 - a_7\phi_f + a_8} \tag{3.49}$$
$$\text{(Exp.: Tsai)}\,,$$

with

$$a_1 = 10534853590571063478476 8513 \,, \tag{3.50}$$
$$a_2 = 469070364718701537167 5802 \,, \tag{3.51}$$
$$a_3 = 10534853590571063478476 8513 \,, \tag{3.52}$$
$$a_4 = 469070364718701537167 5802 \,, \tag{3.53}$$
$$a_5 = 10671742601445530465163 180 \,, \tag{3.54}$$
$$a_6 = 3673775402597027743 830 \,, \tag{3.55}$$
$$a_7 = 11675633988407280502 770 \,, \tag{3.56}$$
$$a_8 = 8358145903237511459 700 \,, \tag{3.57}$$

as well as for the experimental data set by Whitney

Fig. 3.11 Major (lamina) Poisson's ratio as a function of the fiber volume fraction: comparison of experimental values and *optimized* theoretical predictions according to Tsai (**a** experiments by Tsai [2, 3], **b** experiments by Whitney [4])

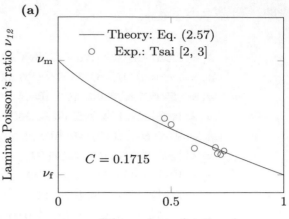

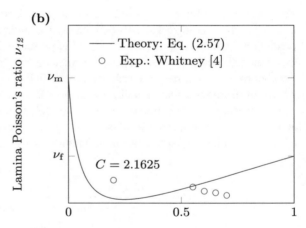

$$G_{12}(\phi_f) = -\frac{(a_1 C + a_2)\,\phi_f^2 + (-a_3 C - a_4)\,\phi_f - a_5}{a_6 \phi_f^2 - a_7 \phi_f x + a_8} \tag{3.58}$$

$$(\text{Exp.: Whitney}),$$

with

$$a_1 = 1246811558802726329728\,, \tag{3.59}$$

$$a_2 = 11069377165147274960\,, \tag{3.60}$$

$$a_3 = 1246811558802726329728\,, \tag{3.61}$$

$$a_4 = 11069377165147274960\,, \tag{3.62}$$

$$a_5 = 22731897742320419745,\tag{3.63}$$
$$a_6 = 7297608310081600,\tag{3.64}$$
$$a_7 = 22151981907792800,\tag{3.65}$$
$$a_8 = 14986252920407700.\tag{3.66}$$

The parameter C (fiber contiguity) must be again adjusted to the corresponding data sets. A graphical comparison between the experimental data sets and the theoretical predictions based on optimized C values is provided in Fig. 3.12.

Fig. 3.12 In-plane shear modulus as a function of the fiber volume fraction: comparison of experimental values and theoretical predictions according to Tsai (**a** experiments by Tsai [2, 3], **b** experiments by Whitney [4])

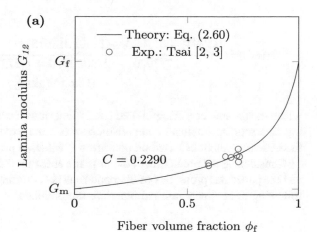

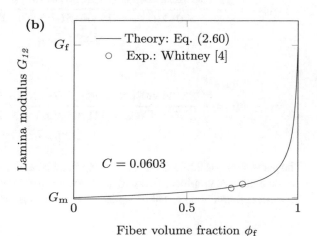

3.3.2 Halpin-Tsai Relationships

Considering the prediction of the modulus of elasticity transverse to the fiber direction according to Eq. (2.63) and the particular properties of fiber and matrix according to Tables 3.1 and 3.6, the following linear/linear rational functions are obtained:

$$E_2(\phi_f) = \frac{(600163347388\phi_f + 29711072161)\,\xi + 629874419549}{8618450\xi - 174092600\phi_f + 182711050} \tag{3.67}$$

$$\text{(Exp.: Tsai)}.$$

$$E_2(\phi_f) = \frac{(1129494001203\phi_f + 11409018615)\,\xi + 1140903019818}{2757900\xi - 273032380\phi_f + 275790280} \tag{3.68}$$

$$\text{(Exp.: Whitney)}.$$

where the parameter ξ (Halpin-Tsai coefficient) must be adjusted to the corresponding data sets. A graphical comparison between the experimental data sets and the theoretical predictions based on optimized ξ values is provided in Fig. 3.13.

Considering the prediction of the in-plane shear modulus according to Eq. (2.64) and the particular properties of fiber and matrix according to Tables 3.1 and 3.6, the following linear/linear rational functions are obtained:

$$G_{12}(\phi_f) = \frac{(19798752361245\phi_f + 881550741187)\,\xi + 20680303102432}{690433605\xi - 15506451675\phi_f + 16196885280} \tag{3.69}$$

$$\text{(Exp.: Tsai)}.$$

$$G_{12}(\phi_f) = \frac{(518313955096\phi_f + 4601667845)\,\xi + 522915622941}{3033700\xi - 341704160\phi_f + 344737860} \tag{3.70}$$

$$\text{(Exp.: Whitney)}.$$

The parameter ξ (Halpin-Tsai coefficient) must be again adjusted to the corresponding data sets. A graphical comparison between the experimental data sets and the theoretical predictions based on optimized ξ values is provided in Fig. 3.14.

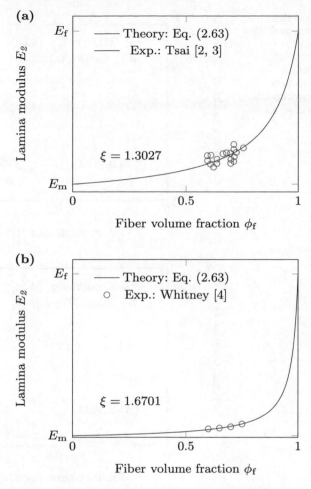

Fig. 3.13 Modulus of elasticity transverse to fiber direction as a function of the fiber volume fraction: comparison of experimental values and *optimized* theoretical predictions according to Halpin-Tsai (**a** experiments by Tsai [2, 3], **b** experiments by Whitney [4])

Fig. 3.14 In-plane shear modulus as a function of the fiber volume fraction: comparison of experimental values and theoretical predictions according to Halpin-Tsai (**a** experiments by Tsai [2, 3], **b** experiments by Whitney [4])

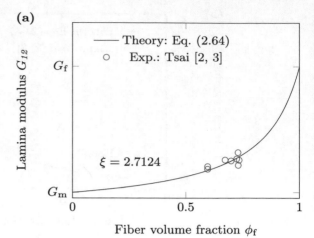

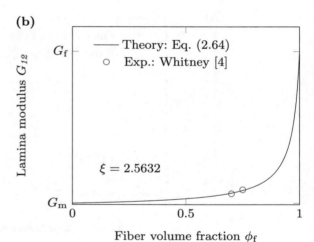

References

1. Press WH, Teukolsky SA, Vetterling WT, Flannery BP (1997) Numerical recipes in Fortran 77: the art of scientific computing (vol 1 of Fortran numerical recipes). Cambridge University Press, Cambridge
2. Tsai SW, Springer GS, Schultz AB (1963) The composite behavior of filament-wound materials. In: Fourteenth International Astronautical Congress, Paris, France, September 1963, Paper No. 139
3. Tsai SW (1964) Structural behavior of composite materials. Technical report NASA-CR-71, Washington
4. Whitney JM, Riley MB (1966) Elastic properties of fiber reinforced composite materials. AIAA J 4:1537–1542
5. Wolberg J (2006) Data analysis using the method of least squares. Springer, Berlin

Printed in the United States
by Baker & Taylor Publisher Services